DOES NATURE APPROVE?

Does Nature Approve?

A Contemplation of Lawn Care Practices

MARK SKELDING

Burlington, Vermont

Copyright © 2020 by Mark Skelding

Cover Photos by Linda P. Keating
Front Cover: Daisy Fleabane (*Erigeron annuus*)
Back Cover: Beardtongue (*Penstemon* 'Dark Towers')

All rights reserved. No part of this publication may be reproduced, distributed, or transmitted in any form or by any means, including photocopying, recording, or other electronic or mechanical methods, without the prior written permission of the publisher, except in the case of brief quotations embodied in critical reviews and certain other noncommercial uses permitted by copyright law.

Onion River Press
191 Bank Street
Burlington, VT 05401

ISBN 978-1-949066-53-1

To my wife, Linda, for her love and the time and support she gave me to do this writing and to my children, Sarah and Sam, for their environmental ethic and appreciation for simplicity and minimalism ... and for being such great kids.

CONTENTS

Foreword
1

~ 1 ~
Pruning
4

~ 2 ~
Weeding
10

~ 3 ~
Weed Trimming
16

~ 4 ~
Inhibiting Weed Growth
24

~ 5 ~
Mulch
32

~ 6 ~
Invasive Plants
36

~ 7 ~
Lawn and Garden Fertilizers
44

~ 8 ~
Raking Leaves, Thatch, and Grass Clippings
48

~ 9 ~
Mowing
52

~ 10 ~
One Yard to Dooryard at a Time
58

~ 11 ~
Addendum: Low Maintenance Perennial Gardens*
62

BIBLIOGRAPHY
70
ACKNOWLEDGEMENTS
72

ABOUT THE AUTHOR
74

FOREWORD

We presume pruning, weeding, mulching, installing weed liners, fertilizing, and spraying pesticides are a necessary part of maintaining our yards. But just how needed and effective are they, and at what cost?

"Beauty" and keeping things "under control" are the primary reasons most of us take these steps. We have accepted that to have a beautiful yard we need to fertilize our lawns to keep them lush, prune our trees and shrubs to keep them "manicured," and get rid of "weeds." And to properly manage our yards we need to prune our trees and shrubs to keep them from becoming "overgrown," spray pesticides to keep away "unwanted" pests, and wage war against weeds to keep them from "overtaking" everything.

Since this is our common perception of what a beautiful and well-managed yard looks like, it follows our standard for property value is the more "well kept" the landscape the higher the value of the home. This, of course, perpetuates these practices.

Understandably, our investment in our homes and our perception of what qualifies as beautiful and tidy influence how we tend our yards. We can't escape our human nature. But there's also nature. *What is nature's definition of beautiful and well-managed?*

The human notions in quotations above reflect our general concept of plants. More importantly they reveal our relationship with nature and how much we are able or willing to accept the wild. All three shape how we garden. *But how does nature garden and do our gardening practices match hers?*

The focus of this book is on common gardening and landscaping practices and their fit with nature. Chapter one focuses on pruning, typically our first outdoor gardening practice of the season.

~ 1 ~

PRUNING

Pruning is the act of removing plant parts. Removing limbs from a tree, cutting back the stems of a shrub, and shearing the outer growth on a plant are examples of pruning. *It's important to recognize that regardless the type of pruning we do, the cuts we make are wounds.*

A plant typically responds to pruning in two ways. First it begins sealing off its injuries because they are potential entry sites for insects and diseases that can weaken and eventually kill the plant. Its second response is to generate new growth to try to compensate for growth that was lost.

Both responses suggest that pruning is not something advantageous to plants. That in fact it's a setback to them. A plant's natural inclination is to continue putting out new growth year after year until the plant naturally senesces (dies of old age). Wounds work to the contrary. They put plants at risk of premature death.

When a plant is wounded its immediate reaction is to wall off its injuries. To do so it seals off the tissue surrounding the wound and basically disconnects that part of the plant from the rest of itself. In essence, the plant divests itself of tissue it put valuable energy into growing; tissue that in turn helped sustain the plant. And to make matters worse, sealing off those wounds requires energy the plant could be putting toward healthy growth.

The plant's second response is to replace its lost growth with a burst of new growth, suggesting that the growth that had been removed was important to the plant. Otherwise, why would the plant have intentionally grown it in the first place, and why would it be worth the energy to try to replace it? The problem with this "replacement" growth is that it's mostly short-term emergency growth such as water sprouts, suckers, and epicormic shoots (multiple stunted shoots sprouting as a single mass on the tree's trunk). This type of growth tries to make up for lost food production for the plant but is not designed for the long haul and in time actually weakens the plant.

So far this sounds like an indictment against pruning. But there's a counter point to all this and that's the suggestion that nature prunes its own. Plants are barraged by things that wound them. Strong winds snap trunks, heavy snow and ice buildup break branches, chewing insects remove foliage, mammals shear twigs, and fungi cause tip dieback. This suggests pruning is not only something natural for plants, it must also be advantageous because plants have been suffering damage for eons yet they continue to thrive.

As we can see, when we try to base decisions on ecology we often end up mired in "cost vs. benefit" and "chicken and egg" scenarios such as we have here. But that's good because it pushes us to think deeply about our interactions with nature.

Do those things that damage plants exist and injure plants for a purpose? Are the injuries they cause actually beneficial to plants both in the short run and for their long term evolution and sustainability?

Or, is it because these damaging agents exist, plants have had to adapt to them? Are the acts of sealing off wounds and replacing lost growth adaptations? If so, are they an evolutionary step forward for plants or are they costly setbacks for them?

If we knew the answers to these questions it would help us know whether pruning is a gardening practice that's right with nature.

Whether we conclude pruning is or isn't part of nature's way, we know one thing for certain. When we prune a plant we wound it. The plant then has to overcome its injuries and will most likely try to compensate for its loss as well. *So if we are going to prune it's important we do so in a way that allows the plant to quickly seal its wounds and replace its lost growth with healthy rather than emergency growth.* The following are some general rules for how to help that happen.

Rule number one is there is no "one size fits all" method of pruning. When, where, and how to prune vary greatly from one type of plant to the next.

When preparing to prune it's important we first make sure our tools are sharp and clean. The sharper the blade the cleaner the cut. This matters because dull blades can tear the plant and tears do not "heal" as quickly as clean cuts. Pruners with blades that overlap generally make the cleanest cuts.

When pruning, tools should be wiped clean with rubbing alcohol preferably after each cut made but definitely before moving on to the next plant. The alcohol helps eliminate bacteria or fungi that may get onto our blades keeping us from spreading disease as we move from one plant to the next.

Proper placement of cuts is critical. Cuts made in the wrong place seal off slowly, if at all, and often become infected by fungi. Generally, cuts need to be made at the junction between the part being removed and the larger section of the plant that part is attached to.

How much can we safely remove from a plant at any one time? The more growth we remove the more stress we put on the plant. The less we remove, the less emergency growth the plant will force and the less energy it will need to invest in sealing off its wounds. The long held rule is to never remove more than one third of a plant in any one season. (Pirone, 1978)

And this final rule brings us back to where we started. What is nature's definition of beautiful and well-managed? *Pruning that, in our eyes, leaves our plants looking beautiful, tidy, and compact may in fact be leaving them for dead. With this in mind and with the jury still out on whether pruning is right with nature, it seems prudent that if we are going to prune we do so conservatively.*

Conservative pruning focuses strictly on the health of a plant and involves two goals. One is to clean out the plant and the other is to increase the ability for air to flow within its canopy.

Cleaning out a plant involves removing dead or broken limbs, limbs growing inward toward the trunk, and any limb that is crossing and rubbing another. Dead and damaged limbs set the plant up for disease and insect attack. Inward growing limbs weaken the structure of the tree and are themselves weak and vulnerable because of the lack of sunlight they receive. And limbs that rub one another result in wounds, potential entry sites for damaging diseases and insects.

Increasing air flow involves selectively removing stems and branches, particularly from the inner portion of the plant. Good air flow is important because it helps keep moisture and humidity from building up inside the plant's canopy. This helps lessen the chance that fungi and bacteria will survive and infect the plant.

But notice! Even if we prune conservatively we're still challenged with our question of whether pruning is right with nature. When we clean out a plant we remove habitat utilized by decomposers, valuable members of the food web. When we thin out a plant's canopy we take valuable cover away from songbirds and alter the environment beneath the plant by increasing the amount of sunlight it's now going to receive. These are just three of countless ripples effected by pruning.

Earlier we asked whether the practice of pruning is ultimately good or bad for plants. Our question has broadened. Is pruning good for nature?

If in the end we decide to go ahead with pruning, this

chapter addresses some general rules about pruning that may help plants quickly seal their wounds and avoid having to force emergency replacement growth. But there are also specific rules that vary from one plant type to the next. The scientific literature and experienced horticulturists and arborists can help with those.

It's often said that pruning is both an art and a science. In a sense it's true. The art is in the eye of the artist and the science is out of respect for the plant.

~ 2 ~

WEEDING

In the last chapter we took a hard look at whether the practice of pruning is right with nature. In this chapter we'll ponder the same about weeding.

First, it's important to recognize that for a plant species to take hold somewhere it has to have a niche. Otherwise it couldn't persist. Having a niche means three things. The species is able to sustain itself with the resources available at that locale. It coexists with the other members of the community living there. And, it coevolves with that community. In other words, the species and its community continually shape and reshape each other for the long term betterment of both in ways impossible to fully predict. (Schoener, 2009)

What does this have to do with weeding? The definition of "weed" is any plant or plant species growing where it's not wanted.

But as we just read, if the plants are growing there they must have a niche. So who wouldn't want them there and why not? Are they unwanted by the insects that feed on them or the pollinators who depend on their flowers? Do the birds that feed on those insects not want those plants around? What about the plants growing alongside them that are benefitting from the nitrogen some of them are fixing in the soil? Of course, these questions presume insects, birds, and plants don't just have needs but wants as well.

Truth be told, it's we who don't want them around. We have come to believe in "the good, the bad, and the ugly" when it comes to plants and we've made up our minds which ones are bad and ugly. We cringe at the sight of dandelions in our lawns, crabgrass in our vegetable gardens, bindweed in our perennial beds, pineapple weed between the stones of our patios, and knotweed in the cracks in our driveways. But those plants have a niche so why not let them serve their purpose? *Because we have accepted our self-created notion of "weed" and the societal norm that weeds detract from the beauty of our property, are a sign of a landscape out of control, and are indicative of a poorly maintained yard.* And so we "weed."

Weeding, unlike weed trimming (see chapter 3), is the attempt to get rid of undesirable plants entirely; all of them, roots and all. We typically try this in one of two ways. We spray them with an herbicide. {*Note: As of 2019, 165 species of weeds had developed resistance to herbicides just here in the United States. It's probably safe to assume the number has risen since then.* (www.weedscience.org., 2020)} Or we uproot them while making sure their seeds and vegetative parts have no chance of producing new plants. *Are there examples of either of these practices in nature?*

With respect to herbicides, yes, there is chemical use

in nature. But are there organisms that use toxins offensively, aggressively attempting to completely rid an area of another species?

Some plants do indeed produce substances that negatively affect other organisms, a process called allelopathy. The roots of walnut (*Juglans sp.*) trees exude juglone which inhibits the growth of certain plants that might potentially compete with them. *Salvia* does the same using camphor. But these are examples of defensive use of chemicals. These substances aren't being used offensively to kill other species, they are simply preventing other species from growing in a particular spots. (Salisbury & Ross, 1978)

Plants that produce repellants are another example of defensive use of chemicals. These substances negatively affect organisms that try to invade or consume the plants. In time those organisms learn to stay clear of these plants.

An example of organisms that use chemicals offensively are the gall producing insects. They inject a substance into their plant host which causes the plant's tissue to grow abnormally, resulting in a gall. These galls serve as protective shelters in which the insects carry out their life cycles. Gall producers use chemicals offensively, but would risk death if they used them to wipe out the very organisms they depend on.

Pathogenic fungi and viruses are also examples of organisms that use substances offensively. One common way plant pathogens do this is by producing enzymes that break down plant cell walls, allowing them entry into the plant and access to the cell contents they require. Plants generally react to these invasions with counter attacks of their own. Although the infected part of the plant may die, the plant itself generally survives. Even if the entire plant does succumb, it's not unusual to see others of its kind nearby and uninfected. In cases where pathogens are able to destroy an entire population it's usually because unnatural conditions (e.g., the species being grown in a monoculture) have increased the population's susceptibility to the pathogen.

It's hard to come up with examples of organisms that *knowingly* use chemicals to kill off entire populations of organisms. That's because all species are interdependent. An organism risks extinction if it destroys species it directly depends upon *and has no reason to offensively go after species it indirectly depends on.*

Uprooting weeds is a second method we employ and, like use of chemicals, uprooting is also found in nature. Squirrels uproot bulbs, deer uproot and consume root crops, and skunks uproot turf in search of grubs. These actions are analogous to weeding, but not in spirit or result. In all these cases random plants are being killed but entire populations are not systematically being eliminated from the community.

Let's pause for a moment and look again at why we weed. Aside from the reasons already stated, we also weed out of concern over *competition*. We don't want weeds robbing sunlight, water, and nutrients from plants we desire. But our concerns about competition may be unfounded. Research is showing that within a plant community there is far more coexistence occurring than there is competition. Where competition primarily exists is between individuals of the *same* species. They not only compete for the same resources but for the same amounts of those resources. That's not the case between populations of different species. In some instances different species require different resources. In cases where they need the same resources, species make trade-offs. Species X requires more of resource A than Species Y, and Species Y requires more of resource B than Species X. (Ackerly, 2009)

What we typically see between different species isn't competition but facilitation. Here are just a few examples. Some species provide shade, helping reduce water loss (transpiration) for others. Some serve as windbreaks, protecting others from water loss and other forms of wind damage. Many species share root systems helping each other to maximize water and nutrient uptake. And all contribute valuable nutrients for one another through the leaf litter they leave behind. Different species affect each other positively which is why they are able to coexist as a community as long as they do. (Callahan, 2009)

Earlier we read that if "weeds" persist they have a niche. So it's hard to justify that weeding them out is nature approved. We also know biodiversity is good. Communities that contain lots of different species with lots of varying traits are healthier than those with fewer species. Wilson, 1992) This makes it doubly hard to justify weeding. But if we do decide to still weed let's be sure the plants we uproot go toward the production of compost. That way what we take away we can at least give back as soil amendment.

This chapter focused on "ordinary weeds" and weeding in its truest sense. The following chapter addresses weed trimming and chapter 6 will focus on a special group of "weeds" known as "invasive weeds."

~ 3 ~

WEED TRIMMING

In the last chapter we questioned whether weeding is right with nature. In this chapter we'll do the same for weed trimming.

When we trim weeds we cut them back either partially or completely to the ground. We typically do this using a hand tool such as a scythe or power trimmers and mowers.

But is nature ok with this? Does nature cut its "weeds" back partially or completely to the ground? Maybe. Partial consumption of weeds by herbivores is a constant in nature and can be compared to the way we cut back weeds. One difference, however, is animals feed on selective plants in a plot and in many cases only on select parts of those plants. Occasionally they may trim a swath of plants within the plot. We typically hack back the entire plot indiscriminately.

Herbivory (grazing) keeps the height of weeds in check but not necessarily their spread. This is because weeds are able to continue propagating despite being chewed on all the time. Running roots that produce new shoots, seeds and multiple means of dispersing those seeds, and new growth to replace consumed growth are just some examples of how they do this. And the reason they're able to do these is because herbivores generally don't harm the plants' roots or crowns. The roots keep the plants alive and their crowns keep them growing and spreading.

Weeds being cut down to ground level also occurs in nature. Grazing animals clip plants close to the ground, somewhat analogous to the way we mow. Like "trimmer herbivores," grazers consume the above ground portions of plants but not the plants' crowns and root systems.

It takes special plants to withstand grazing. Forbs, such as plantain, dandelion, and curly dock, are a good example. They have basal rosettes which are leaves that radiate from the plants' crowns but are hard for grazers to get at because they lie flat on the ground. Forbs typically have taproots as well. These deep, sturdy roots anchor the plants and help keep them from being uprooted by grazers. They also store food to compensate for leaf loss. And, since grazing increases the soil's exposure to wind and sun, these roots help ensure the plants have access to water despite dry surface conditions.

Grazing is comparable to mowing. However, grazers in the wild do not "mow down" every plant across an entire plot and they don't graze the exact same spot over and over again like we do when we mow patches of weeds we don't want growing "out of control."

This quick look at herbivory suggests three things for us to consider when it comes to trimming weeds:

- It's not natural for an entire plot of weeds to be cut back.
- It's not natural for the same individuals to be trimmed over and over again.
- Individual weeds and weed species survive despite being trimmed.

All three suggest it is nature's intent for "weeds" to persist. They appear meant to have a niche and contribute to their community's biodiversity.

Their niche is not only how they sustain themselves where they are but how they facilitate the other organisms there. This might include providing shade for plants that can't tolerate full sunlight or dry soil, drawing soil water closer to the surface for plants with shallow roots to access, fixing nitrogen in the soil making it more available for others, or taking up certain nutrients that might be nearing toxic levels for some plants.

These facilitative effects enable coexisting plants to persist and may create conditions for new species to move in. This enhances the biodiversity of the community. Weeds contribute to biodiversity in other ways as well. They are protective cover for small mammals and food for herbivores. They are refuge for various insects, including beneficial parasites and predators. Their seeds are eaten by birds and other animals. Their flowers help sustain certain pollinators. The dead plant material they leave behind each year is food for decomposers and other members of the soil community.

So knowing all this, do the practices of cutting back or

mowing down weeds fit with nature? Generally, we trim weeds in hope of containing their spread, reducing habitat for "pest" species such as rodents, improving our view of what lies beyond them, and giving our yards a more "tidy" appearance. Do any of these simulate natural herbivory and impact weeds and weed communities the way herbivores do?

Cutting back weeds won't stop them from spreading by way of their root systems and neither will mowing. However, mowing can inhibit or alter the growth of shoots coming from those roots. It may deter shoots from sprouting where they continually get knocked back or it may cause the shoots to change their growth habit in a way that protects them from the mower. For example, their shoots might grow more horizontally than before and their leaves may be smaller.

Trimming can, however, prevent plants from spreading by seed. If plants are cut back or mowed as their flowers begin to die this will halt seed production or at least prevent seeds from fully developing.

Regularly mowing weeds down to ground level will drastically reduce protective cover for rodents. But it reduces habitat for the other members of the community as well. And with mowing comes soil compaction and the creation of conditions that only certain soil organisms and plants can tolerate. What results with repeated mowing is a gradual, unhealthy decrease in biodiversity.

Simply cutting weeds back can also reduce cover depending on the species of plants growing there. Some plants may produce replacement growth more dense (protective) than what they had before they were cut back.

Mowing weeds down may improve our view of our overall property. Just cutting them back can as well, though granted to a lesser degree. Knowing that mowing has a greater ecological impact on our yards than simply cutting plants back, we're left with a trade-off. How much view are we willing to sacrifice for the benefit of our yard as a whole?

And finally, tidy appearance. Some view weeds as unsightly, others see them as beautiful wildflowers. Knowing that weeds have a niche and contribute to biodiversity we're left with a question similar to the one regarding view of our property. How much "untidiness" are we willing to tolerate in exchange for the benefits weeds provide?

Whether we mow weeds down or just cut them back, the natural community at large is hurt less than if we completely weed them out. But as we've seen, both practices have their drawbacks and there's one more worth mentioning. When we indiscriminately cut back weeds we are essentially shearing them. The cuts we make are wounds and the plants will respond by putting energy into healing those wounds and producing replacement growth. The lower we make our cuts the larger the wound we leave and the more foliage (food production for the plant) we remove. In other words, the lower the cut the more we set the plant back.

The effects of this setback go beyond just the weeds themselves. Less foliage means less food for herbivores and less refuge for beneficial insects. Energy directed at healing injuries and replacing lost foliage means less energy going into flower and eventual seed production. This hurts pollinators and seed eaters (dispersers).

Our quick look at herbivory suggests some things to consider when trimming weeds and this brief look at plant ecology and physiology does, too.

If we are going to mow or cut plants back we should do so selectively. Trim just some plants, not all of them, and be sure to spare individuals of every species present.

Trim as infrequently as possible. Less frequent trimming puts less stress on the plants and allows them more time to rebound to their natural state. It's also less upsetting to the rhythm of *natural* disturbances the larger community is already undergoing. {*Note: The more disturbed an area, the more vulnerable it is to invasive species.*} (Chisholm, 2009)

Trim once the plants' flowers start to die back. This will prevent the plants from producing viable seed which will help slow their spread. Although this will hurt seed eaters at least pollinators can still benefit.

Avoid mowing. If that's not possible, mow high. Mowing is actually a form of plant inhibition. The adaptive pressure it puts on plants, the soil compaction it leaves behind, and the increased soil exposure to sun and wind it allows lead to conditions that exclude many species from being able to grow there. In short, mowing reduces biodiversity. *And ironically, the plants that mowing favors are typically the very weeds people don't want.*

And finally, take as little off the plants as possible.

Higher cuts are less of a physical setback for plants and, in turn, less of a setback for the community as a whole.

The bottom line in all this? We know that if weeds persist there's an ecological reason why. We also know weeds can and do persist despite being trimmed. And we know there are limits to how, how much, and how often weeds can or should be cut back or mowed.

~ 4 ~

INHIBITING WEED GROWTH

Weed inhibition is the fourth gardening practice we'll investigate. To inhibit weeds means to stop them from spreading, suppress their growth, or prevent their seeds from germinating in the first place.

There are four strategies we commonly use to try to inhibit weeds. One is to place black plastic over an area we want weed free. The heat that can build up under the plastic and lack of sunlight work together to kill the above ground portion of existing plants and prohibit new plants from growing.

A second strategy, used most often for perennial beds, is to use a weed barrier. This is done one of two ways. One is to cover the bed with a single liner of mesh before perennials are planted and then cut holes in the liner where perennials can be placed. A second way is to plant perennials first and then place separate sheets of mesh around them. This method is less effective than the first because weeds inevitably find the seams between sheets. Once the liner and perennials are in place a layer of mulch is spread on top of the liner for aesthetic purposes and in hopes of maintaining soil moisture and further suppressing weeds.

Cultivation is a third strategy. Periodically disturbing the surface of the soil by lightly scratching through it with a cultivator prevents plants from becoming established.

The fourth strategy is to plant groundcover or perennials in hopes that these plants, or at least the shade they produce, will exclude weeds from growing there.

Are there processes in nature comparable to any of these four strategies? Let's start the comparison by looking at our use of black plastic, weed mesh, and mulch. Our primary purpose with all three is to *cover the soil to prevent weeds from growing.*

Nature covers its soil, too. During the growing season plants blanket the ground. The plants' foliage protects the soil from drying out from overexposure to sun and wind. It also protects it from wind erosion. The plants' roots help maintain healthy soil structure by aerating the soil and stabilizing it against water erosion.

In autumn, as leaves fall and plants die back, the ground is littered with a layer of plant debris. This layer helps maintain soil moisture and moderate soil temperature, helping the plants' roots withstand winter conditions. As spring approaches decomposition of this layer quickens. The litter is returned to the soil as nutrients and organic matter and the plants coming up through the litter take their turn once again at blanketing the soil.

An important trait of plant cover and plant litter is their permeability. Their porosity allows growing plants to push up through them and falling seeds to reach the soil below them. It also allows rainfall and snowmelt to infiltrate the soil. And thanks to this porosity the soil can "breathe." Gas exchange between the air and soil is important for it ensures healthy soil and is the primary way atmospheric nitrogen, a critical nutrient for plants, enters the soil.

Notice the striking difference between our use of soil covers and nature's. We use them to try to *discourage* plant growth and nature uses them to *encourage* plant growth. In fact, nature's primary cover *is* plants.

Black plastic is essentially impermeable. But water and gases can still move from adjoining soil into soil that's covered with plastic. This means two things. One, roots of plants can and will still be present under the plastic. And two, conditions under the center-most section of the plastic can be very hot, very humid, low in gas exchange which can result in anaerobic conditions favoring certain fungi and bacteria.

Is black plastic effective at stopping weeds? It shuts out sunlight and this prevents above ground portions of plants from growing. However, plastic does not stop roots from growing. Survey the edge of the plastic and you will see weeds growing out from under it. Lift the plastic and you'll see a network of roots meandering their way across the surface of the soil in search of "greener pastures." Remove the plastic and before you know it the weeds are back. Leave the plastic on season after season and it will gradually breakdown, allowing weeds to break through.

To show our black plastic truly worked we would have to remove it, remove all surface and below ground roots that have been growing under it, immediately plant what we'd like to have growing there, and then hope those plants thrive and the weeds stay away. But if that were to work can we know for sure it was the plastic that did the job?

Unlike plastic, weed mesh *is* porous. This is one of the reasons why it fails to prohibit weeds. It may delay weeds from emerging, but it doesn't exclude them. In fact, as weeds work their way through the mesh their roots and the mesh become entwined. This actually ends up protecting the weeds. We see just how protected they are when we try to weed them out and discover we can't because they're enmeshed.

When mesh is installed in perennial beds it isn't just weeds it delays. Mesh also slows new shoots growing from the perennials' crowns and roots. If the perennials were free to flourish and spread perhaps weeds would be less numerous.

And finally, mulch. Mulch can also be porous depending on its composition and depth. Mulch typically consists primarily of woody material such as shredded brush and bark. Because the goal of using it is to stop weeds, it is usually spread at depths of three or four inches or more. Under certain conditions the top layer of mulch can become hard and tightly packed. This crust can prohibit some weeds from breaking through. It can repel water as well.

Though mulch and plant litter might seem comparable, they're not. Their composition and function differ. Both consist of cellulose but not to the same degree. Plant litter is comprised of material not just from woody plants but herbaceous plants, too. Like mulch, plant litter blankets the ground but rarely do we find litter as thick as the mulch in our gardens. Plant litter is permeable. Mulch can be too but we use it in hopes it won't be. Litter decomposes quickly. Mulch decomposes slowly. Litter encourages plant growth. We hope our mulch will discourage plant growth. *In short, nature litters but it doesn't mulch.*

Mulch is very popular with gardeners and landscapers and, like anything, it has its positives and negatives. In the following chapter we'll look closely at the pros and cons of using mulch.

Cultivation is the third strategy listed at the top of this chapter. By periodically cultivating the surface of the soil we can inhibit plants from taking root. But cultivation will not work unless a thorough weeding of the area takes place first (see chapter 2). Otherwise, cultivation will simply wound the plants that are there. They'll survive their injuries and may come back even stronger than before (see chapter 3).

If the area is thoroughly weeded roots and all, cultivation will work. Cultivation brings seeds buried in the soil to the surface, drying them out and interrupting stratification and germination. Any seedlings that do happen to sprout are uprooted by cultivation. When a seedling is uprooted its root hairs are torn and this usually kills the seedling. Cultivation discourages roots of other plants from encroaching into the cultivated area. And, cultivation keeps the soil exposed. This allows soil to dry out making it difficult for invading seeds to germinate.

Soil exposure helps inhibit weeds but it's a critical downside of cultivation. Exposed soil is vulnerable to erosion and prolonged exposure can alter the soil's chemistry, composition, and structure. Not surprisingly, nature keeps its soil from becoming exposed. If soil is exposed nature acts quickly to cover it back up. In both cases nature's approach is the same. It covers its soil with a layer of plants. And yes, included among those plants are weeds.

Strategy four is to plant ground cover and perennials. This approach to trying to inhibit weeds is exciting because it's actually somewhat comparable to what we see in the wild.

In nature we find plant communities comprised of diverse species coexisting, or at least co-occurring. Plants that are coexisting are typically in facilitative relationships. They help sustain one another. For example, the shade one species creates protects another. Sometimes, however, competitive relationships will develop and disrupt that coexistence. If conditions are right, one or more species can gradually exclude others by creating conditions that keep them out. Or they may displace others by creating conditions that force them out. Shade is again a good example. It can both exclude and displace species.

Planting ground cover or perennials can work to exclude or displace weeds. This is especially true if the plants we put in are a) locally common, b) planted where they can persist, and c) as diverse a mix of species as possible. One important distinction, however, between how we typically implement this strategy and how nature does it is time. We tend to do mass weeding and planting for immediate results while natural succession of plants is gradual.

And, even if we take steps a-c, success is not guaranteed. We can never know for sure that what we plant will persist (i.e., if they're the "right plants in the right place"). There are so many permutations of environmental factors governing the ebb and flow of natural communities it's impossible to accurately predict which plants will persist. And how do we know the different species we plant won't end up developing competitive relationships? If they do it could open the door for weeds to return. And ... how do we know for sure the plants we're putting in are actually going to exclude or displace weeds? It's possible they may end up facilitating them.

So once again ecology has left us with more questions than answers. But that's a good thing. It makes us think more deeply about our interactions with nature.

Our efforts against weeds are never ending. There has to be a reason for that. Is it because our methods are inadequate? Is it that our reasons for wanting weeds gone are contrary to nature's way? Or, is it simply nature telling us "weeds" *are* the "right plants in the right place?"

~ 5 ~

MULCH

Our practice of trying to inhibit weeds was last chapter's topic and one inhibition method we looked at briefly was mulching. In this chapter we'll look more deeply into the pros and cons of using mulch.

One takeaway from the last chapter is that nature litters but it doesn't mulch. As much as we may hope that our use of mulch simulates nature's use of plant litter, it's a challenging case to make. For starters, plant litter encourages plant growth while mulch is meant to discourage plants from growing. But let's look further.

The pros and cons listed below may provide additional help as we wrestle with whether mulching is nature approved. Lists like this remind us that for every positive there's a negative and that a single trait can be advantageous and disadvantageous at the same time. Such is the case with mulch.

Pros	Cons
Aesthetically pleasing	Expensive
	Unsightly as it decomposes/washes away
Delays weeds	Does not eliminate weeds
Prevents soil pathogens splashing on leaves	Can come with weeds, weed seed, and fungi
	Can "incubate" weeds and fungi
Can protect/harbor beneficial insects	Can protect/harbor harmful insects
Can decompose slowly	Can decompose quickly
Typically effective for an entire season	Typically effective for only one season
Its breakdown nourishes the plants	Its breakdown nourishes weeds
	Breakdown of woody mulch robs soil of nitrogen
Can be permeable to water and gas exchange	Can be impermeable to water and gas exchange

Can be impermeable to weeds	Can be permeable to weeds
Can help conserve soil moisture	Can prevent rain from reaching the soil or adequate soil depth
	Can encourage shallow rooting
Prevents wind erosion of topsoil	Inhibits aeration of soil underneath
Helps moderate soil temperature	Can decay tree bark it's in contact with
	Provides cover for stem/trunk girdling rodents

These are just some of the arguments for and against mulch. No doubt there are others we could add to the list.

There's one final topic to address on mulching. In the last chapter we focused on mulch used on perennial beds. Mulch is often spread around the trunks of ornamental trees as well.

The three most common reasons given for spreading mulch around the bases of trees are 1) it's aesthetically pleasing, 2) it takes away any reason for getting mowers and weed trimmers close to trunks, possibly injuring them, and 3) it helps preserve soil moisture for the trees' roots.

The first reason listed is subjective, of course. The second is sound and if trees could talk it's a good bet they'd thank us for it. The third, however, is worth another look. The roots that are most actively taking in water are far from the trees' trunks. In fact, most are well beyond the tips of the trees' most outer branches. This and some of the cons listed above, particularly tree bark decay, suggest mulching around trees may be more costly than beneficial. Protection from power tools and some of the pros listed above suggest the opposite.

These chapters are written in hopes of arousing our ecological consciences, fueling conversation, and perhaps stirring a little constructive controversy. The following chapter on invasive species promises to do all three.

~ 6 ~

INVASIVE PLANTS

The four chapters prior to this pertained to "ordinary weeds" … everyday plants we commonly find growing in our lawns and gardens. Dandelion, chickweed, pigweed, plantain, and crabgrass are some examples. There is also a group of "weeds" many consider inordinate. They're known as "invasive weeds." In this chapter we will explore the notion of "invasive" plants and whether our disdain for them is justified and our fight against them is right with nature.

An invasive plant is defined as any exotic (non-native) species that takes over an area in a relatively short period of time at the expense of the other species living there. There are two critical things we've learned about invasive plants. One is that disturbed soils are most vulnerable to them. And two, they're able to take over rapidly because, being non-native, there are no coevolved enemies present to keep them in check. (Chisholm, 2009)

The term "invasive" accentuates the speed and domination with which these plants become established. It also connotes badness. So to put that in proper context we need to also consider invading (as opposed to invasive) plants and aggressive plants.

Every plant and every plant species is an invader. Otherwise they wouldn't exist. Wind, water, animals, and gravity help the plants' seeds invade new territory. Their roots, rhizomes, and stolons invade nearby soil.

All plants are aggressive as well. Again, they have to be if they're to hold their own. Some are more aggressive than others. Brambles, morning glory, mint, Virginia creeper, wild grape, and garden phlox are examples of species that can be very aggressive.

There are two schools of thought regarding invasive species. One is that they don't belong where they've invaded and are bad because they reduce native plant populations and local biodiversity. The second acknowledges the impact they have on natives and biodiversity, but goes on to remind us of how ecological processes work over time. Adaptation, coevolution, coexistence, and facilitation will eventually arise even with monocultures of invasive species. Though they are monocultures they still interact with the larger plant and animal community surrounding them. That interaction will gradually generate ecological relationships between them and their larger community. Like man, no monoculture is an island.

To further stir the invasive species debate let's consider the following notions.

- It's not just exotics that are "invasive" (i.e., rapidly take over an area at the expense of the

other organisms living there). Native plants can also take over areas and displace their fellow natives.
- It's arguable that every native species had to first be an "invasive" ... or at least an aggressive exotic.
- A species can be invasive in one locale, aggressive in another, and simply coexist in a third. So is it really the particular species that's the problem?
- There are many non-native species that coexist with native species and are not invasive. Again, is it the species that's "bad?" Or does the problem lie with the habitat that's being invaded?
- Extinctions that have resulted from invasive species have been primarily in the animal kingdom and by way of disease or predation, not competition for resources.
- Human mobility, global exchange, and environmental disturbance are what enable a species to become invasive. *If humans are part of nature then how can their impacts, including introduction of invasive species, not be natural as well*? (Chisholm, 2009)

So with these thoughts in mind, should we be trying to weed out or at least inhibit invasive species? Those of the first school of thought on invasive species will likely answer yes. Those of the second school will say, "Not so fast."

We know that in the short term invasive plants displace native plants and reduce local biodiversity. *What we don't know is what their long term consequences might be.*

Given our certainty about their immediate effect and uncertainty about their future impact, it makes sense our first step should be to try to turn away any new, potentially invasive species until we know more about their long term effects.

Federal and state plant quarantines are one approach to doing that. Initially, quarantines were mostly a reactionary step, with plants being quarantined after being discovered to be invasive or noxious. Quarantines are now being used proactively as well. This is challenging, though, because there are so many factors that determine if and where an exotic plant might become invasive. Quarantines will help slow the introduction of invasive plants, but plants spread in ways other than human transport.

A second proactive step also focuses on human activity. Land that has been or is continually disturbed is vulnerable to invasive plants. If we allow land to remain in its natural state it will be less apt to give way to invasive weeds. And if we restore land we've disturbed, we help make it harder for invasive weeds to get a foothold there. We can never restore land to its exact pre-disturbed condition, but we can revitalize it by replanting it with a diverse mix of native plants adapted to the site's new conditions.

Preventing exotics from becoming established is one thing, but what about invasive plants that are already here? Our choices are to try to weed them out, repeatedly trim them back, try to inhibit them, or let nature take its course.

Chapter 2 addressed weeding and whether nature uproots its plants or uses herbicides to eradicate them. The answer to both seems to be no ... at least for "ordinary weeds." Should invasive plants be an exception to that rule?

The environmental hazards of herbicides and the ability of organisms to develop resistance to them are well documented. Despite this, herbicides are being used in cases where there's concern that the potential costs of the invasive plant outweigh the potential costs of using herbicides. In some of those cases it has worked. Most success stories, however, resulted from an integrated pest management (IPM) approach which involved a combination of mechanical (e.g., uprooting the invasive weed), cultural (manipulating site conditions to disfavor the invasive weed), biological (bringing in natural enemies of the invasive weed), and chemical controls. These efforts take many years and many dollars to achieve success. And what if that success unravels once IPM ends? (Murdoch, 2009)

Uprooting the plants is a second approach to trying to "weed" out a species. Success cases involving uprooting are very hard to find. It's hard and very time consuming work. Because the plants have no natural enemies they are able to quickly take over large areas which means there are a lot of them to have to dig out. Having no enemies also means they can devote more energy into growing strong root systems and less energy into defending and healing themselves. Anyone who has tried to uproot a plot of goutweed, creeping bellflower, or Japanese knotweed knows about unrelenting root systems. Because these plants typically take advantage of disturbed soils, this may also explain their prolific root systems. And one final challenge with uprooting weeds, including invasive ones ... even the smallest plant fragment left behind can generate new sprouts.

If we decide against trying to weed them out completely, our next two choices are to either continually cut them back or try to inhibit their growth and spread. The plusses and minuses of trimming plants back versus mowing them down were the topic of chapter 3. Chapters 4 and 5 discuss the costs and benefits of trying to inhibit weeds from growing in the first place using black plastic, weed liners, mulch, and cultivation.

A fifth type of inhibition mentioned in chapter 4 is to grow certain plants that will exclude or displace unwanted plants. Of all the approaches toward "weeds" we've looked at, this approach seems most right with nature. Might it work with invasive weeds? Are there native plants we could bring in that are adapted to the site that's been taken over by invasive weeds and aggressive enough to co-occur, and perhaps someday coexist, with them (see chapter 2)?

Our last choice when it comes to what to do with already established invasive weeds is to simply let them be and let nature take its course. And this brings us back to our two schools of thought regarding invasive species. Some points in the debate were mentioned earlier in this chapter. Here are some follow up thoughts for what promises to be a long debate on whether our battle against invasive species is nature approved.

- Invasive weeds like garlic mustard, goutweed, and Japanese knotweed reduce abundance of native plants and local biodiversity. Thousands upon thousands of acres of soybeans, wheat, and other *exotic* agricultural crops do, too.
- The environmental and economic cost of producing those agricultural crops is astronomical. Is that because monocultures go against nature's design so to sustain them requires monumental cost? *Might it also be because they are exotics and given time exotics lose their invincibility?* Remember, invasive species are exotic.
- Soybeans are just one of many exotics that have proven valuable. Given time, might invasive species eventually prove valuable?
- *Humans* are enabling native plants to become exotic invasive plants elsewhere. The plants are simply responding to the opportunity being handed them.
- If a native or exotic plant species is able to persist somewhere that means it has a niche and is serving some purpose at that location (see chapter 2).
- *Invasive species do not wipe out native species, they displace them.*
- Can we know for sure that the pressure invasive plants put on native plants won't in time be a force that evolves native plants for the better?

The topic of invasive species is contentious and can rile the emotions. Hopefully this chapter simply stirs contemplation. In the following chapter we'll take a look at fertilizer use.

~ 7 ~

LAWN AND GARDEN FERTILIZERS

I debated on whether to include something here in lieu of what was to be the chapter on this practice or to just skip the topic altogether. It is such an important subject I decided I had to at least encourage the reader to consider it along with the other practices addressed in this book.

In 2016, *Horticulture Magazine* purchased this essay on lawn and garden fertilizers separate from the series of essays (now chapters in this book) it was part of. That essay would have been this chapter. It is on file at www.hortmag.com.

The essay addresses the following question. *Is use of commercial lawn and garden fertilizers nature approved?*

Clearly, nature fertilizes its soil and does so in a sustainable way. For without sustained soil fertility natural plant communities could not exist. So if nature is already naturally fertilizing its own soil why are we applying additional fertilizer to our lawns and gardens?

The essay compares nutrient cycles, particularly nitrogen and phosphorus, to residential use of commercial fertilizer. It also looks at cultivation practices that will reduce, if not eliminate, the need for supplemental fertilization, particularly practices that increase biodiversity. Examples of how biodiversity enhances a plant community's ability to *self*-fertilize include:

- Greater diversity of plants leads to greater diversity of organic matter and decomposers to help fuel nutrient cycles.
- More species of plants means more layers of plants within the community. The more layers there are from ground level to canopy, the more protected the soil is from erosion (soil loss = nutrient loss) and exposure (drying out = less soil moisture = less soil solution = less available nutrients).
- Greater diversity leads to greater facilitative relationships between plants. Facilitation sustains plant diversity and abundance (see chapter 2). More plants equal more nutrients and organic matter ending up back in the soil when those plants die and are decomposed.
- Diversity facilitates more nutrient sharing tradeoffs between plants, which helps slow soil depletion of various nutrients.
- Different species of plants have different types and depths of root systems. Roots aerate the soil and greater aeration means better gas exchange with the air, water and nutrient percolation and

availability, and more suitable habitat for a greater variety of decomposers ... all critical to nutrient cycles.

Is residential use of commercial fertilizers nature approved? After all, nature fertilizes so fertilization must be nature approved, right?

I hope the reader is able to access the full essay. If not, I hope this brief summary helps stimulate thoughtful consideration over whether to use commercial lawn and garden fertilizers.

~ 8 ~

RAKING LEAVES, THATCH, AND GRASS CLIPPINGS

Does nature rake up and dispose of plant debris? Observe forests, meadows, and even regularly mowed lawns and it's easy to see the answer is no. Wind sometimes blows debris around leaving a place look as if it's been raked. But wind only relocates plant debris, it doesn't remove it from the system. The plant debris still ends up becoming part of the local soil as nutrients and organic matter.

Of all the gardening practices we've critiqued so far, raking and disposing of leaves, thatch (accumulated grass remnants in turf), and grass clippings is perhaps most contrary to nature. Yet it's the most common large scale practice done to lawns, next to fertilizing and mowing.

If nature doesn't rake, why do we? The main reason we rake is mentioned in the foreword of this book. A lawn that's full of leaves or covered with grass clippings is "messy."

There's a second reason we rake and it, too, is driven by human nature. If plant debris is left on our lawn, it can shade out grass leaving thin or "dead" spots in our turf. It can also become habitat for "pests" and disease that might feed on or infect our grass. But as we read in chapters 4 and 7, nature's way is to create and then use its plant debris. If we were to take our lead from nature we'd let plant debris do its thing. But doing so would leave our lawns "blemished" and "unsightly."

Unlike the other practices we've looked at, the ecology behind this one is relatively cut and dry. We know that removing plant debris from a plant community robs it of the very nutrients and organic matter it needs to sustain itself. But we also know that if we allow plant debris to accumulate and then decompose in our lawns season after season, plants that are adapted to habitats where that naturally occurs will gradually establish themselves. In time, our "perfect" carpet of grass will look more like a field (mowed meadow).

Most homeowners want the "perfect" carpet of grass. But "perfect" lawns are essentially monocultures, something nature is not. In fact, nature works tirelessly to rid itself of monocultures. This is why trying to maintain monocultures requires such an inordinate amount of human intervention.

So where does this leave us? We can let nature and plant debris take their course ... and our turf along with them. Or we can rake and then try to compensate for lost nutrients and organic matter with supplemental watering, fertilizing, reseeding, aeration, and weed control. But the direct and indirect economic and environmental costs of doing the latter are substantial.

There might be one practice worth considering, though, and that's topdressing our lawns with compost. Composting has its critics. They argue the economic and environmental costs of compost production more than offset the benefits of composting, basically defeating the purpose. They also point out that even though composting restores carbon to the soil, it also releases carbon into the atmosphere and we should be investing in solutions that *reduce* carbon emissions.

But home composting may be a compromise when it comes to raking. What if we manually rake up leaves, thatch, and grass clippings, compost them, and then spread the compost where we raked? Our carbon footprint will be smaller, we'll get some exercise, our soil will be replenished, and our lawns will be lush and "tidy" all at the same time. And ... if enough of us do this then Scotty from Scotts won't have to constantly yell at us to "Feed it!"

The topic of the next chapter is mowing. We read in chapter 3 how mowing is somewhat comparable to herbivory ... at first glance. But the more we read, the less support we found for the practice.

~ 9 ~

MOWING

We read in chapter 3 that some believe mowing is simply mechanical grazing. A key difference between mowing and grazing, however, is how regularly lawns are mowed compared to how often a plant or even plant community is grazed. Grazers feed selectively over a large area, moving from one spot to another, eating parts of plants here and there. When we mow we cut every plant down close to the ground and we do that to the same plants in the same spot over and over again. Although a grazing animal's range is typically larger than our lawns, which process ultimately "consumes" more biomass, mowing or natural grazing?

There's another natural phenomenon some may compare to mowing. Wildfires indiscriminately reduce every plant over a large area down to ground level. But wildfires don't occur weekly.

It's hard to defend that mowing is nature approved. For starters, one of the main reasons we mow is so our lawns won't transform into meadows. But trying to thwart plant succession and keep biodiversity from increasing is as contrary to nature as we can be ... aside from outright destroying it.

The following chart lists some examples of the dichotomy between nature and mowing.

Nature	Mowing
Plant growth is indeterminate (plants continually grow)	Mowing prevents natural growth habit
Plants shade soil, keeping it moist	Mowing takes away plant shade
Plants protect soil from wind exposure	Mowing removes plant cover
Exposed soils are vulnerable to invasive species	Mowing removes plant cover
Plant litter preserves soil moisture and moderates soil temperature	Mowing grinds up potential plant litter
Plant litter becomes soil nutrients and organic matter	Mowing reduces plant biomass
Healthy soil is permeable and full of air pockets	Mowing compacts soil
Plants take in atmospheric carbon (carbon dioxide)	Burning mower fuel adds carbon to the air

Plants remove carbon from the atmosphere	Mowing reduces plant biomass
Nature is habitat	Mowing reduces habitat
More habitat = healthier, more complex food webs	Mowing reduces habitat
Nature is biodiverse	Mowing reduces biodiversity
More biodiversity = a more sustainable balance of specialist vs. generalist organisms and more complex food webs	Mowing reduces biodiversity
Our "weeds" are nature's wildflowers	Mowing creates conditions that favor "weeds"

Mowing seems to go against nature. There is one possible argument in defense of mowing, though, and it relates to the first thing listed in the chart. Perhaps mowing has actually strengthened the overall fitness of those plants that have adapted to it. Mowing has forced them to practice phenotypic plasticity. They have fine-tuned their ability to alter their form in response to environmental stress. For example, plants that under normal conditions grow vertically can, when subjected to frequent mowing, switch to growing low to the ground. Being able to adapt to environmental stress is an advantage and perhaps mowing enhances that ability for some plants. (Travis, 2009)

For the most part, however, mowing is unnatural. But it's hard to imagine homeowners abandoning mowing. Keeping our lawns mowed is one of our social norms. Neatly trimmed grass is pleasing to the eye and a carpet of turf is comfortable to walk and play on. Mowing reduces habitat for insect pests that annoy us while we're trying to spend time in our backyards. It also reduces cover for rodents that might damage or invade our homes. Composting mowers help return grass clippings and leaves to our soil. Those who use push mowers get some exercise. And those using manual reel mowers get even more exercise, at the same time doing something to help reduce their carbon footprint.

These are the benefits of mowing. But beneficial to whom? The only one in the list that appears beneficial to nature is composting mowers that grind up plant debris and quicken its return to the soil. But how much of a true benefit is that knowing how much more nutrient cycling there would be if our yards were allowed to grow?

Is there a compromise? Can we better abide by nature yet still appease our human nature?

There are two options that would allow us to do that. Both begin with reducing the amount of lawn we mow. Option one is to simply allow those areas no longer being mowed to naturally transform from lawn to meadow. Down the road when brambles, shrubs, or tree saplings start to move in, if we don't want plant succession to go beyond meadow, we can do some minimal mowing and uprooting to help our wildflowers remain dominant.

The second option is to plant perennials in those areas

no longer being mowed. This option is more costly, labor intensive, and perennial gardens are generally less biodiverse than meadows. But with perennial gardens we can choose which plants we want and after a couple of years of adding new plants and dividing and transplanting existing plants, our perennials will dominate and exclude or displace "weeds." Viola! No mowing, no weeding, and no raking. We would have beautiful foliage and flowers to enjoy and our yards would be one step closer to being natural plant and animal communities. (See Addendum)

Our yards would also be a step closer to being true yards. They would be more like dooryards.

If mowing can't be avoided here are some suggestions to help make it a little more ecological. Mow as infrequently as possible. Mow as high as possible. Never mow lower than three inches. Use as light a mower as possible to help reduce plant damage and soil compaction. Use mowers that don't produce carbon emissions. Mow in a different pattern each time to help reduce soil compaction and plant damage. Don't mow during hot, dry periods to avoid "burning" the grass and leaving soil exposed. Be sure blades are sharp so plants are wounded as little as possible.

~ 10 ~

ONE YARD TO DOORYARD AT A TIME

This book has challenged us to take a hard look at our gardening practices. Are they in line with nature? The previous chapters have helped us see that for the most part they are not. In fact, it seems if anything our gardening relationship with nature is simply a battle of wills.

For instance, plants continually add new growth each growing season. We constantly cut that growth back with our saws, pruners, weed trimmers, and mowers.

Over time the species that make up a plant community change. Eventually, the community itself changes. We, on the other hand, do everything possible to keep our lawns and gardens looking essentially the same year in and year out.

Nature tends toward greater and greater plant diversity. We manipulate our lawns and gardens so that only a select few get to grow there.

Nature litters and then uses its debris to fertilize its soil. We remove nature's litter and then "rob Peter to pay

Paul," compensating for the loss using commercial fertilizer.

Nature protects its soil from exposure. We open it up to exposure through low mowing and raking.

And though we know these things, we're probably not going to stop controlling weeds, mowing, raking, and maintaining our turf anytime soon. Why? Human nature. As we read in the foreword, we like order and feeling a sense of control. We tend to obey social norms and respond to peer pressure. And yes, of course we appreciate the wild and all its beauty... as long as it isn't closing in on us.

But what if we were to give in a little? What if there's a compromise that will better align us with nature yet still appease our human nature?

Over the years we have adopted a notion of "yard" that's actually opposite the original yard. Up close to our houses we plant shrubs and perennial gardens, and away from our houses we clear and mow vast expanses of lawn. What if we reversed that? What if we went back to the days of the dooryard? Convert our lawns into perennial gardens or allow them to revert to meadow, and clear and mow a relatively smaller area up close to our homes.

It sounds like a radical proposal. And it is ... literally. The term radical comes from the Latin word for "root." A return to dooryards would be a return to our roots. It would eliminate the need for just about all of the gardening practices discussed in this book ... practices that have essentially enslaved us.

We'd have far less mowing or weed trimming to do. We

would no longer need to do any pruning because trees and shrubs would be away from our houses and left to grow naturally. There would be no need for us to uproot weeds or spray them with herbicides. We wouldn't need to inhibit weed growth, either, so we'd no longer have any use for black plastic, weed liner, or mulch. Fertilizer would no longer be required because our yards would be healthy plant communities that would cycle their own nutrients. Raking would be obsolete which would further fuel those nutrient cycles. And finally, eliminating all of this human disturbance to our yards would greatly lessen the chance that invasive plants could establish themselves there.

If we were to do this, we wouldn't be going *back* to the days of the dooryard, we'd be moving *forward*.

~ 11 ~

ADDENDUM: LOW MAINTENANCE PERENNIAL GARDENS*

It takes a lot of effort to establish a new perennial bed and the labor can be back breaking. And that's just the beginning, right? Don't perennial gardens require a ton of work to maintain?

Whether our garden ends up high or low maintenance depends primarily on two things ... our notion of what qualifies as a perennial garden and our acceptance of plant population dynamics. So before we talk about the "how to" of low maintenance perennial gardening let's talk philosophy.

What exactly is a perennial garden? One that contains perennials, of course, but there can be many aspects defining such a garden. They may include size, number of plant varieties, variety of flower and foliage color, range of foliage texture, and a seasonal continuum of flowering. Range of plant height and layering, number of bird and pollinator attractors, range of habitat-providing plants, variety of edible plants, and ratio of seeders to spreaders may also apply. And of course there's ratio of natives to exotics. *Which aspects we consider integral to perennial gardens dictates how much work we make for ourselves.*

How much we're willing to accept the notion that plant populations naturally fluctuate and allow that to happen impacts our work load as well. The ecology of a plant community is governed by a complex web of interdependencies that exists among its members. These relationships are in constant flux and forever evolving. As they change, the community's makeup shifts. *The variety of species in a community and the number of individuals within those species inevitably change over time.*

How accepting are we of species "coming and going" in our perennial gardens? Are we fine with plant succession and content letting nature be nature? Or do we prize certain species and do whatever it takes to preserve them?

And this brings us full circle. *What we accept as a "perennial garden" is the garden that results from our intervention with the plants growing there.* Our perennial gardens (notion one) are simply manifestations of our level of acceptance/tolerance of species "coming and going" (notion two). Since, by default, our stance on notion two takes care of notion one, let's focus on notion two.

For those of us willing to let nature take its course, our work is easy. Once we plant our initial garden we simply sit back and watch it evolve. Perhaps we don't even bother with initial planting. Maybe we simply remove the sod where we want our garden and allow wildflowers to naturally take over there. Or, maybe we don't even bother with sod removal. Perhaps we simply stop mowing that area and let it gradually transform to a meadow.

Since nature tends toward greater biodiversity, if we let nature do its thing the aspects of a perennial garden listed earlier will all be met with the possible exception of one. Depending on our surrounding plant community we could end up with a garden consisting of native plants only. Otherwise, with an increase in variety of species comes a wider range of flower color, flowering time, foliage color and texture, plant heights, etc. *When we let nature take its course it both designs and maintains our perennial gardens.*

It's a different story for those of us uneasy with "letting go and letting nature." Many of us fall into this category. And those of us in this group typically share four thoughts regarding perennial gardens.

First, we expect some degree of permanence. We have our favorite plants and we plant them in hopes of enjoying them year after year.

Second, human intervention is necessary. For without weeding, deadheading, fertilizing, seed pod removal, pruning, thinning, dividing, and transplanting our plants will eventually succumb to the forces of nature.

Third, we desire, and perhaps prefer, exotic plants. Native plants are generally too dull and commonplace for us. In fact, to some of us they're simply "weeds." Exotics, on the other hand, are unique and showy. It's not a perennial garden, we feel, if it doesn't contain exotics.

And fourth is our leaning toward exhibit gardens rather than seamless, meadow-like gardens. Although nature may be orderly in its own way, it's not what we typically consider neat and aesthetically pleasing. We want our plants displayed and free of entanglement from other plants, and we want our clusters of individual species distinct and homogenous.

Of course, all four of these conflict with how nature works. And going against nature means work ... hard, persistent work. Again, the aspects of a perennial garden we choose to focus on and how much we want them featured determines how much labor our gardens will demand. Compound that with loyalty to these four beliefs and we're guaranteed hours of work in the garden.

And for many of us that's just fine. Working in the garden is fun, therapeutic, gets us out outside, and provides us with some exercise. It's a labor of love. But for those who wouldn't mind less time *in* the garden and more time sipping their favorite beverage *beside* the garden, here are five things to consider when designing, preparing, and planting our gardens.

First, start small. Because flowers are so pleasing it's natural we want gardens everywhere. But establishing them requires considerable labor. If we take on more than we can handle we could end up overwhelmed and frustrated.

Once we've determined what's manageable our next step is to prepare the site for planting. To help ensure few weeds as possible repopulate our garden it's critical we don't simply rototill the site. Rototilling churns up weed seed and vegetative plant parts but doesn't rid the soil of them. It also aerates the soil for those seeds and plant parts. Instead, if we remove all sod and dig out any remaining weeds, root and all, we will have better results.

Once our site is prepared it's time to plant. Plant selection is critical. Most importantly we need to choose plants we know are the right plants in the right places. Matching the plants' growing requirements with characteristics of our site such as amount of sun, soil moisture, soil pH, and winter temperatures will help ensure this. Not surprisingly our best matches are generally our already persisting native plants.

With our list of possibilities now narrowed there comes a second round of plant selection. *For our garden to be low maintenance the plants we choose have to be able to quickly claim territory and maintain both themselves and the garden as a whole.*

Broadleaf perennials such as hosta, lady's mantle, and large-leaved varieties of cranesbill fill a garden quickly and produce plenty of shade that helps exclude weeds. If leaves become too expansive they can easily be clipped off at their base. *Perennials with dense crowns* and thickly packed stems exclude weeds from growing among them and fill gaps steadily without overtaking neighboring plants. Examples of such plants include Solomon's seal, phlox, and bee balm. And finally, *prolific seed producers* such as globe thistle, lupine, and jewelweed rapidly fill gardens as well. If too many seedlings germinate, or seedlings arise in places they're not wanted, a single weeding soon after germination easily keeps spread of the species in check.

These three types of plants are all low maintenance

plants and when planted together their combined growth habits eliminate much of the work typical in maintaining a perennial garden. Their quick expansion reduces our need to continually plant and transplant in order to fill gaps. They're aggressive enough to stake their claim but not aggressive enough to displace one another. This reduces our need to divide or extract plants. And ... they exclude weeds. For many of us, weeding is highest on our list of high maintenance tasks and our least preferred chore.

A fifth thing to consider is use of mulch and landscape fabric. Yes, they help retain soil moisture but neither prevents weed growth. They simply delay weeds from emerging. Once weeds start emerging mulch makes it extra difficult to extract them root and all. Landscape fabric poses an even bigger challenge. Many weeds can grow up through the fabric and many can root down through it. And with fabric, the roots become entangled with the material which makes removing weeds root and all impossible.

Contrary to weeds, many perennials are unable to sprout up or root down through landscape fabric. Consequently, their crowns are restricted from expanding and their shoots are unable to poke through.

Mulch and landscape fabric have their place, but not in perennial gardens. Our best solution for retaining soil moisture and avoiding weeds is to plant the three types of perennials mentioned earlier and commit to some initial cultivation, our next topic.

If we go with these five suggestions our low maintenance garden will practically be accomplished and we'll barely be into our first season.

Our only labor from here on will be three simple tasks. The first is cultivation. Second is the once-a-season job of expanding the garden's border. And third is the annual job of removing dead plant material.

Cultivation is relatively effortless especially in healthy soil. Using a three or four prong long-handled cultivator we simply scratch up the soil to perhaps an inch deep. This tears up germinating weeds and deters roots, rhizomes, and stolons of unwanted plants from creeping into our garden. It also aerates the soil which benefits our perennials.

Initially we will need to cultivate around our newly planted perennials to uproot weeds trying to fill any open space. We will also have to cultivate along the edge of our bed to repel encroaching sod and plants such as crabgrass and ground ivy. Once our perennials fill the garden the only cultivation required will be along the border of the garden. That will only be necessary every three to four weeks if sod removal is part of our initial preparation.

Our second task is to expand our garden's border each spring once the ground thaws. This will enable our perennials' crowns to continue to expand. We'll remove a six to eight inch band of sod and then cultivate the new edge as previously described.

Finally, our third task is annual removal of dead plant material. This is commonly done in the fall but is that the best time? If a significant number of our plants end up diseased it's wise to remove that debris as early as possible. Otherwise, there are benefits to waiting until early spring. Insects that overwinter in stalks and seeds left in pods are food for birds. Uneaten seeds are eventually dispersed and may produce additional plants come spring. Dead plant material provides extra insulation for our perennials' roots during the winter. Finally, the breakdown of that material returns nutrients and organic matter to the soil, reducing the need for fertilizer.

So where does all this leave us? Even if we can't bring ourselves to "let go and let nature" it's still possible to have a relatively low maintenance perennial garden if that's what we desire.

*I am extremely grateful to the University of Vermont Master Gardening program, Burlington, VT, for publishing an excerpt of this chapter in *Gardener to Gardener* (Vol. 9, No. 4; Fall, 2016).

BIBLIOGRAPHY

Levin, S. A., Editor. (2009). *The Princeton Guide to Ecology*. Princeton, NJ: Princeton University Press.

- Ackerly, D. D. Phylogenetics and Comparative Method. I.16.
- Callahan, R. M. Facilitation and the Organization of Plant Communities. 4.
- Chisholm, R. The Ecology, Economics, and Management of Alien Invasive Species. 8.
- Houghton, F. A. Terrestrial Carbon and Biogeochemical Cycles. 11.
- Murdoch, W. Biological Control: Theory and Practice. 1.
- Schoener, T. W. Ecological Niche. 1.
- Travis, J. Phenotypic Plasticity. 9.
- Vitousek, P. M. & Matson, P. A. Nutrient Cycling and Biogeochemisty. 10.

Pirone, P. P. (1978). *Tree Maintenance*. New York, NY: Oxford University Press.

Salisbury, F. B. & Ross, C. W. (1978). *Plant Physiology.* Belmont, CA: Wadsworth Publishing Company.

Statista Research Department Report. www.statista.com. April 15, 2019.

The International Survey of Herbicide Resistant Weeds. www.weedscience.org. January 30, 2020

Wilson, E. O. (1992). *The Diversity of Life.* Cambridge, MA: Belknap Press/Harvard University Press.

ACKNOWLEDGEMENTS

I gratefully acknowledge the following individuals for their roles in the evolution of this book.

My clients over the years have been invaluable and I thank them on two accounts. Thank you for having the courage to consider ecology first rather than economics or aesthetics when deciding how to tend your lawns and gardens. And second, thanks for being willing to take the risk and allow me to expand my lifelong action research in plant ecology beyond just my own backyard.

I thank the great ecologists and environmentalists, particularly Rachel Carson, Aldo Leopold, and Robert Van Den Bosch, for inspiring me to be political on the topic of ecological correctness.

I'm especially grateful to the Deep Ecologists, in particular Arne Naess and Fritjof Capra, who were integral in my discovery that ecology, for me, was not just an interest, fascination, and passion but actually something spiritual; a way of being.

Thank you so much Rachel Fisher, Publishing Manager and Onion River Press for making this book a reality.

And finally, I have special gratitude for my mother,

Shirley Miriam Lenhart Skelding. Despite her lasting fear as a Great Depression survivor she supported me in my desire to be a horticulturist even though she was sure it would impoverish me. Thank you, Mom. I love you.

ABOUT THE AUTHOR

Mark Skelding has been a groundskeeper, landscaper, nurseryman, arborist, and greenhouse manager. He has a Bachelor of Science degree in horticulture from Pennsylvania State University and was a graduate research assistant in entomology at Michigan State University. While there he served as a cooperative extension intern before accepting an entomologist position at a local arboretum.

He has written articles for journals, magazines, and newsletters, as well as authoring a weekly newspaper column. He also developed a graduate level ecological literacy/schoolyard habitat course, accredited by Southern New Hampshire University, which he taught for six years to hundreds of educators around the state of Vermont.

Most recently he was proprietor of Green with Envy, a small niche perennial garden, shrub, and tree care business in St. Albans, Vermont where he resides with his wife, Linda, continues his garden ecology research, and goes fly fishing every chance he gets.

www.ingramcontent.com/pod-product-compliance
Lightning Source LLC
Chambersburg PA
CBHW060032040426
42333CB00042B/2402